I like to eat these 3.

I don't like these 2.

- Can count to 3.
- Can write numerals to 4.

Notes/date:

How many?	
eyes	
arms	
feet	
spots	
hairs	
ears	
other	

- Fairly confident with early numbers.
- Can recite numbers to
- One to one correspondence

Notes/date:

1.3

three three

Make 3 of each.

What can you find 3 of?
Draw them.

- Can talk about work.
- Can make up to 3

Notes/date:

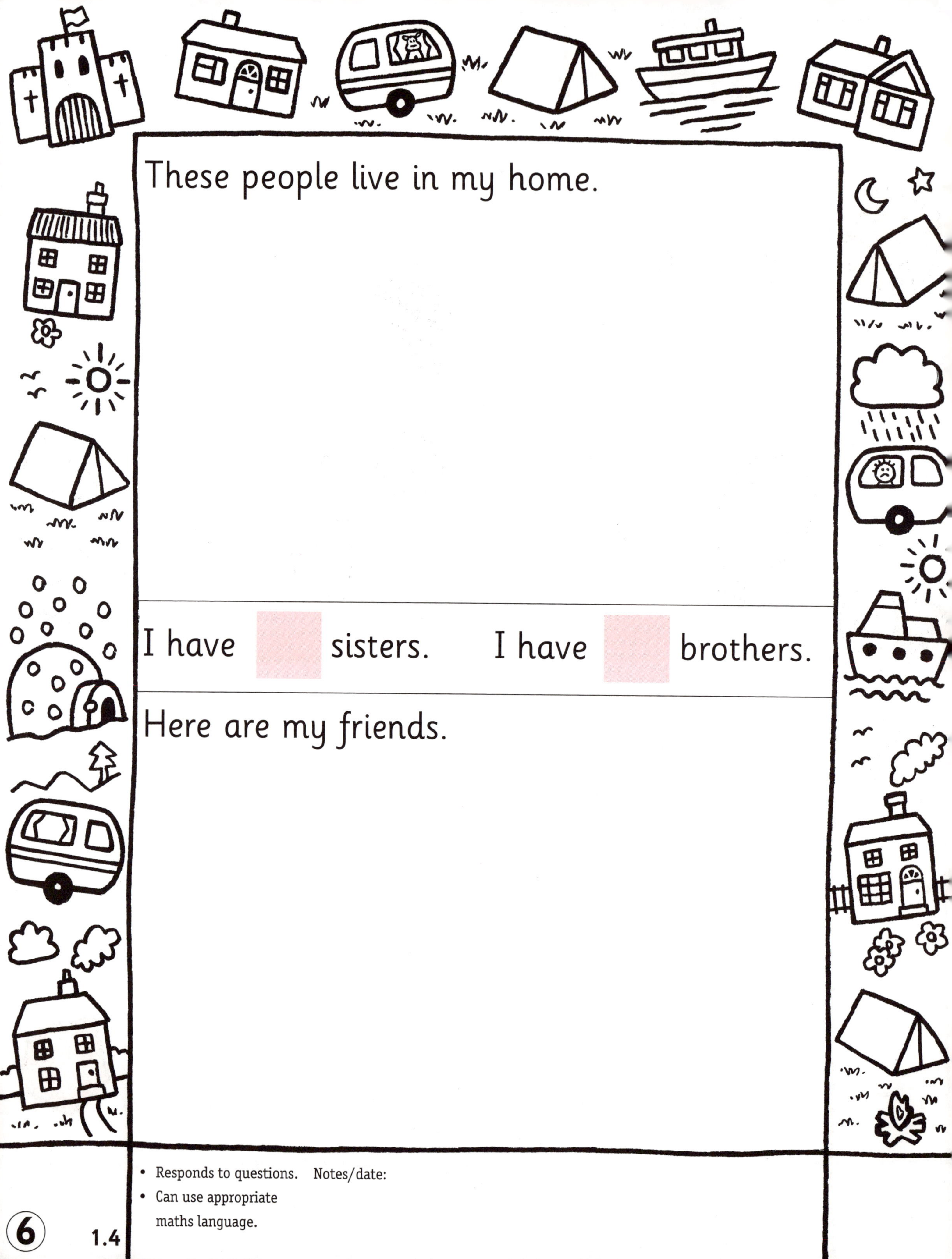

These people live in my home.

I have ☐ sisters. I have ☐ brothers.

Here are my friends.

- Responds to questions.
- Can use appropriate maths language.

Notes/date:

• Can describe a route. Notes/date:

Draw things you made with playdough.

How many?

What else did you make?

How many?

- Can use appropriate language for counting.

Notes/date:

Give 1 teddy a balloon.

Colour 2 teddies brown.

Give 3 teddies a hat.

Give 4 teddies a box.

Give 5 teddies an orange.

- Knows last number in the count is the number in the set.
- Uses language of counting more/less than.

Notes/date

- Can write numerals.
- Can order numbers to 10.

Notes/date:

My biggest number is

The smallest is

- Asks questions about work.
- Understands larger/ smaller numbers

Notes/date:

. . . . holds more than . . . .

. . . . holds more than . . . .

. . . . holds  

. . . . holds  

- Can use language of comparison Notes/date: (more than/less than)

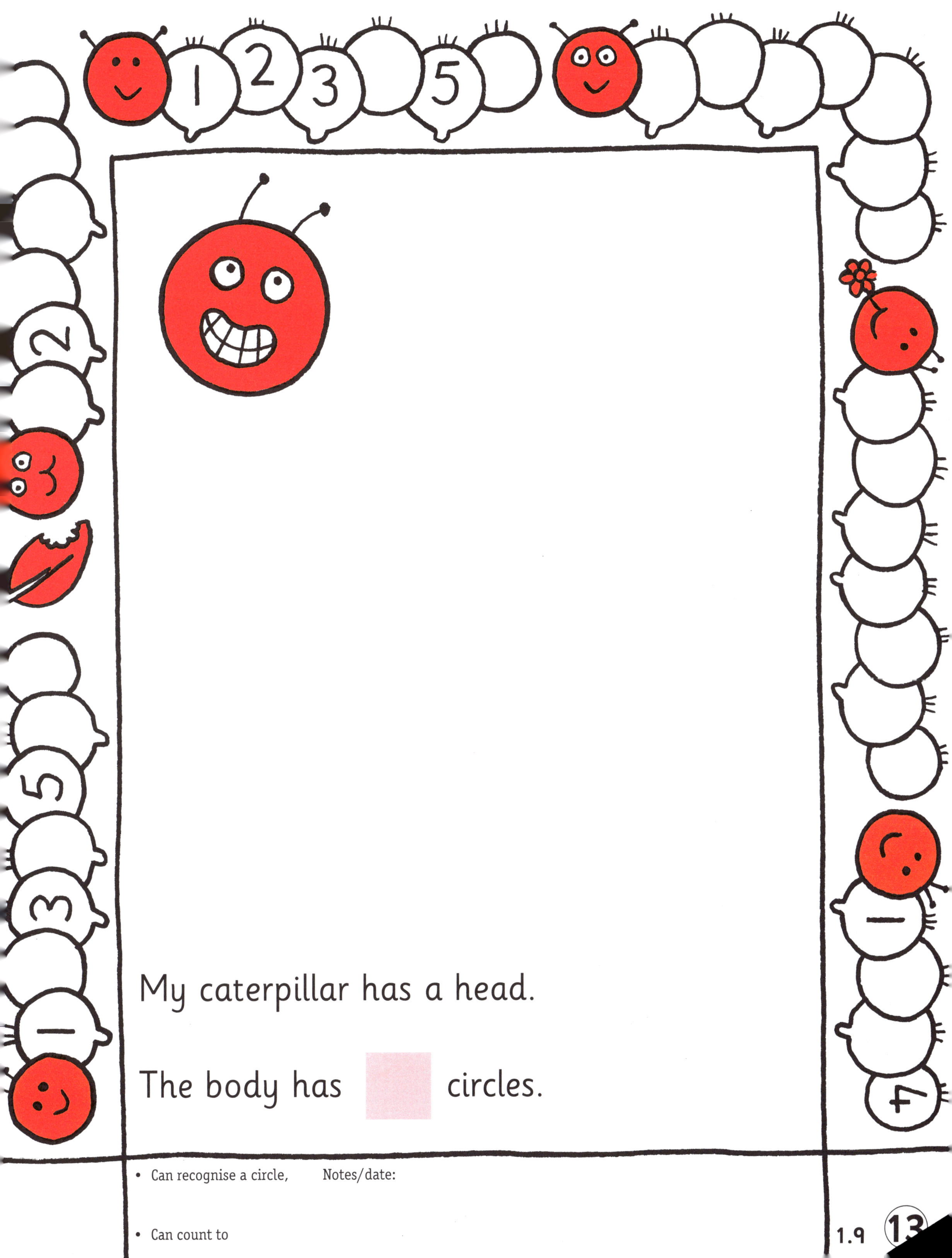

My caterpillar has a head.

The body has circles.

- Can recognise a circle,
- Can count to

Notes/date:

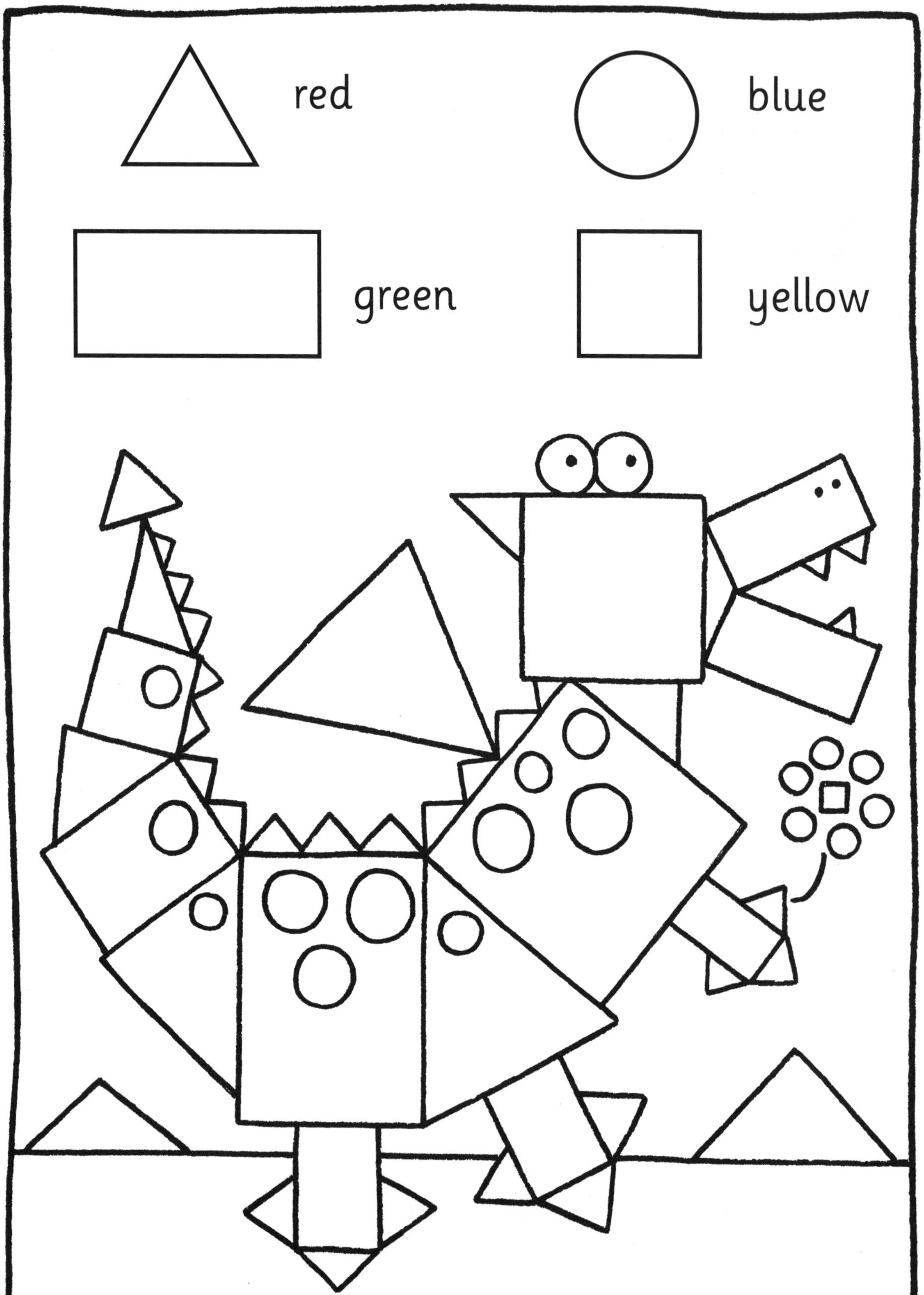

red

blue

green

yellow

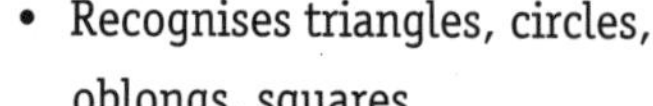

- Recognises triangles, circles, oblongs, squares
- Can count to

Notes/date:

I can name
these shapes.
date
I like best
in maths
date
I can
count to
date

PUBLISHED BY THE PRESS SYNDICATE OF THE UNIVERSITY OF CAMBRIDGE
The Pitt Building, Trumpington Street, Cambridge CB2 1RP, United Kingdom
CAMBRIDGE UNIVERSITY PRESS
The Edinburgh Building, Cambridge CB2 2RU, United Kingdom
40 West 20th Street, New York, NY 10011–4211, USA
10 Stamford Road, Oakleigh, Melbourne 3166, Australia

Sue Atkinson Sharon Harrison
Lynne McClure Donna Williams

Illustrated by Cathy Baxter

First published 1995
Third printing 1997

Printed in the United Kingdom at the University Press, Cambridge

CAMBRIDGE
UNIVERSITY PRESS

ISBN 0-521-47581-3
9 780521 475815